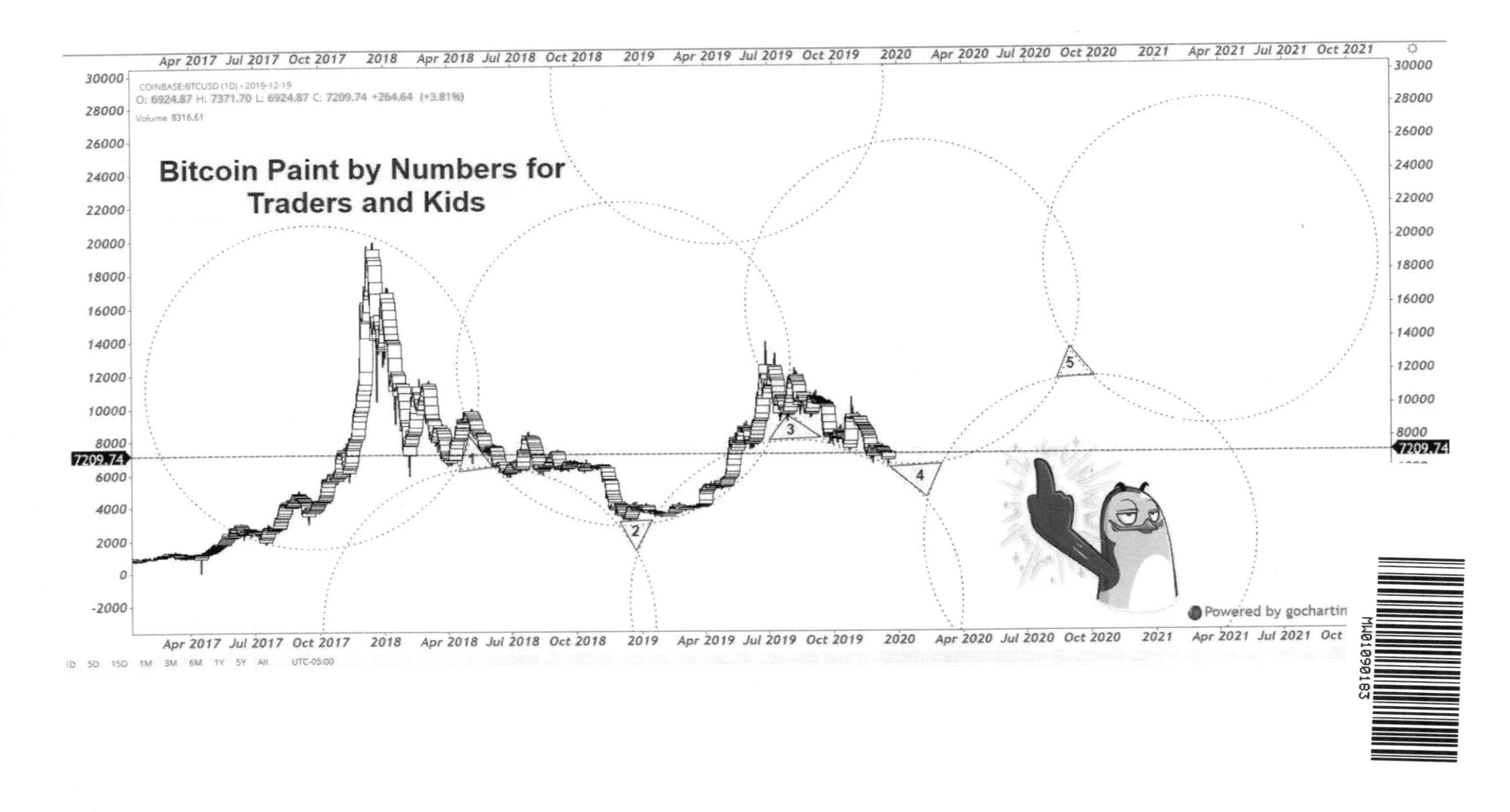
Bitcoin Paint by Numbers for Traders and Kids
COINBASE:BTCUSD (1D) · 2019-12-19
O: 6924.87 H: 7371.70 L: 6924.87 C: 7209.74 +264.64 (+3.81%)
Volume 8316.61
Apr 2017 Jul 2017 Oct 2017 2018 Apr 2018 Jul 2018 Oct 2018 2019 Apr 2019 Jul 2019 Oct 2019 2020 Apr 2020 Jul 2020 Oct 2020 2021 Apr 2021 Jul 2021 Oct 2021
30000 28000 26000 24000 22000 20000 18000 16000 14000 12000 10000 8000 6000 4000 2000 0 -2000
7209.74
1
2
3
4
5
Powered by gochartin
1D 5D 15D 1M 3M 6M 1Y 5Y All UTC-05:00

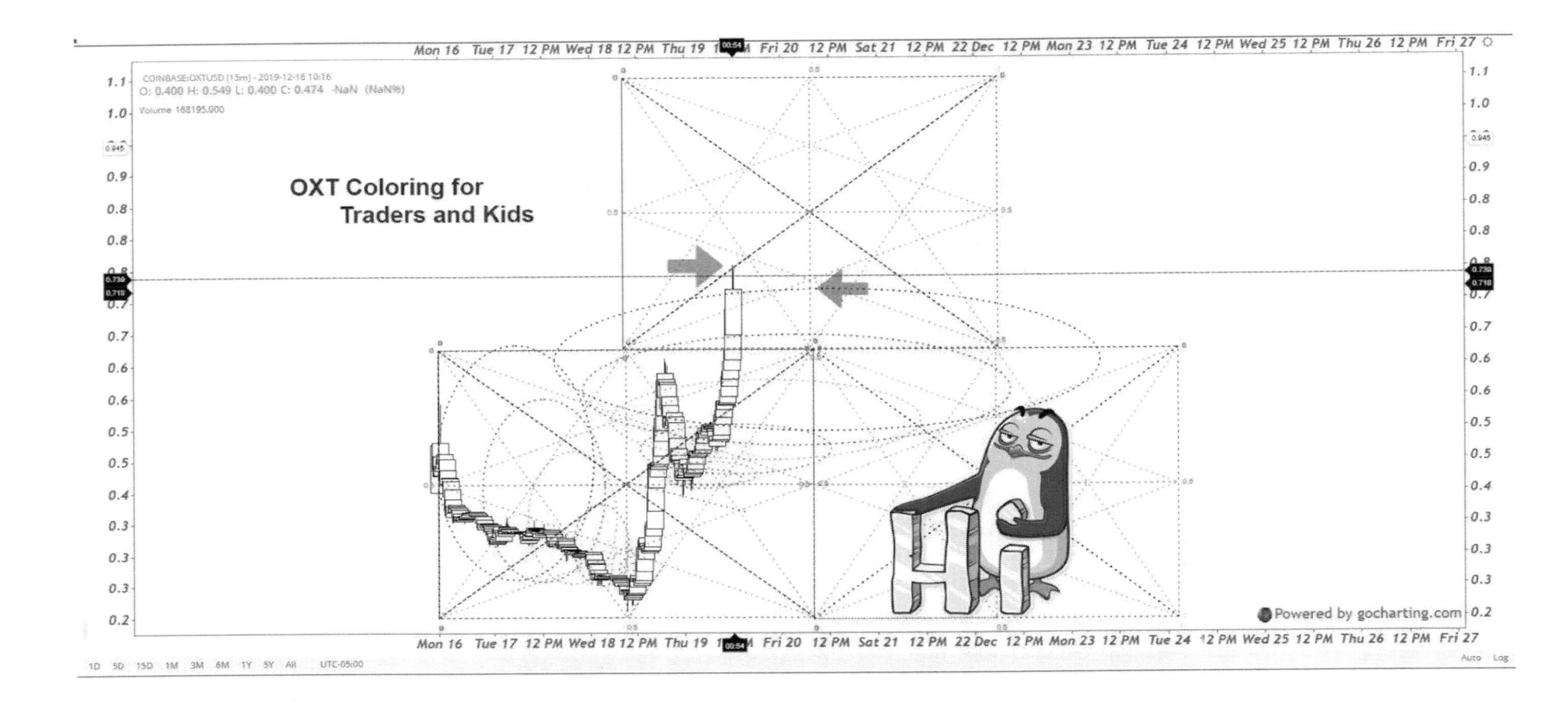
OXT Coloring for
Traders and Kids
COINBASE:OXTUSD (15m) - 2019-12-16 10:16
O: 0.400 H: 0.549 L: 0.400 C: 0.474 -NaN (NaN%)
Volume 168195.000
Mon 16 Tue 17 12 PM Wed 18 12 PM Thu 19 Fri 20 12 PM Sat 21 12 PM 22 Dec 12 PM Mon 23 12 PM Tue 24 12 PM Wed 25 12 PM Thu 26 12 PM Fri 27
Powered by gocharting.com
1D 5D 15D 1M 3M 6M 1Y 5Y All UTC-05:00
Auto Log

COINBASE:ETHUSD
1D
Jul 2017 Oct 2017 2018 Apr 2018 Jul 2018 Oct 2018 2019 Apr 2019 Jul 2019 Oct 2019 2020 Apr 2020 Jul 2020 Oct 2020
440 420 400 380 360 340 320 300 280 260 240 220 200 180 160 140 120 100 80 60 40
127.63
1D 5D 15D 1M 3M 6M 1Y 5Y All
UTC-05:00
Auto Log

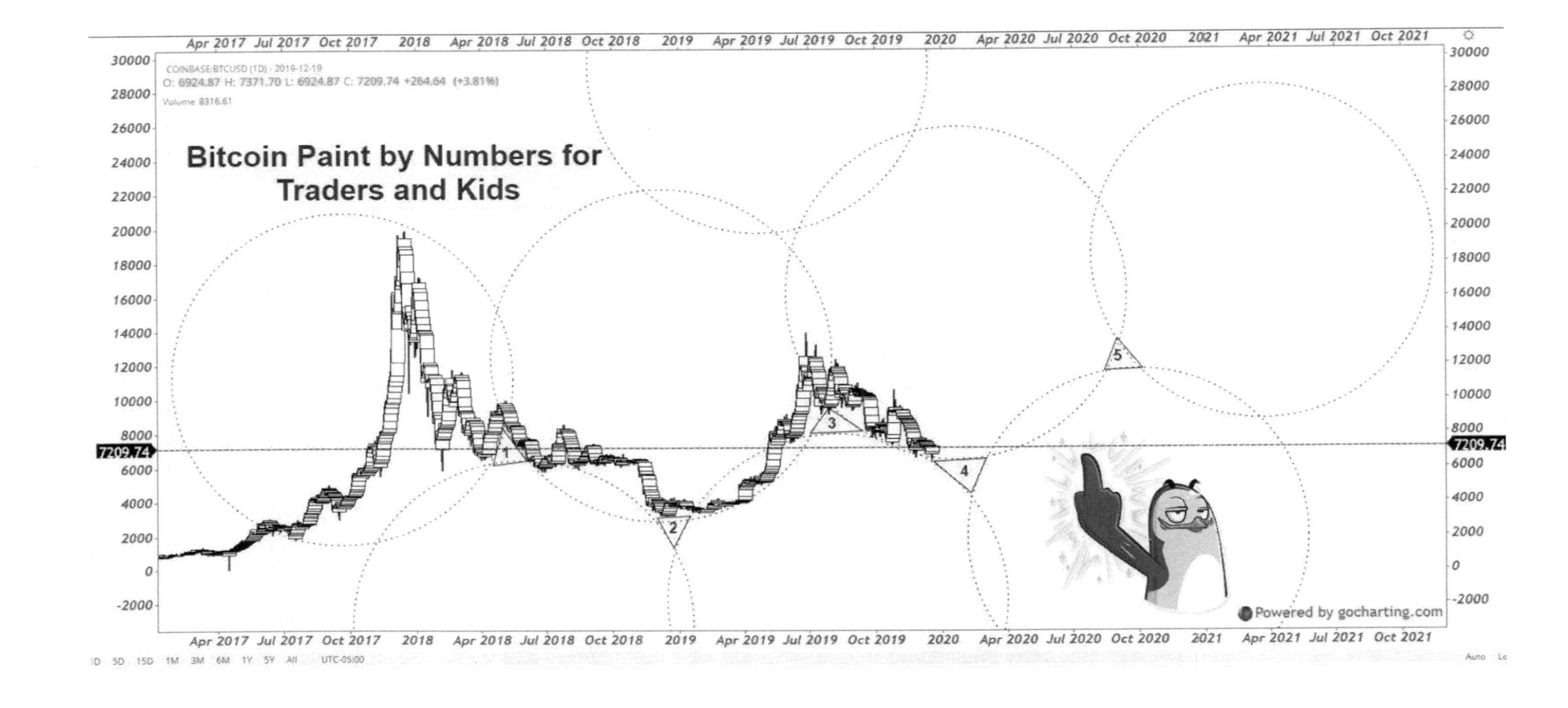
Bitcoin Paint by Numbers for Traders and Kids
COINBASE:BTCUSD (1D) · 2019-12-19
O: 6924.87 H: 7371.70 L: 6924.87 C: 7209.74 +264.64 (+3.81%)
Volume 8316.61
Apr 2017 Jul 2017 Oct 2017 2018 Apr 2018 Jul 2018 Oct 2018 2019 Apr 2019 Jul 2019 Oct 2019 2020 Apr 2020 Jul 2020 Oct 2020 2021 Apr 2021 Jul 2021 Oct 2021
30000 28000 26000 24000 22000 20000 18000 16000 14000 12000 10000 8000 7209.74 6000 4000 2000 0 -2000
1
2
3
4
5
Powered by gocharting.com
1D 5D 15D 1M 3M 6M 1Y 5Y All UTC-05:00

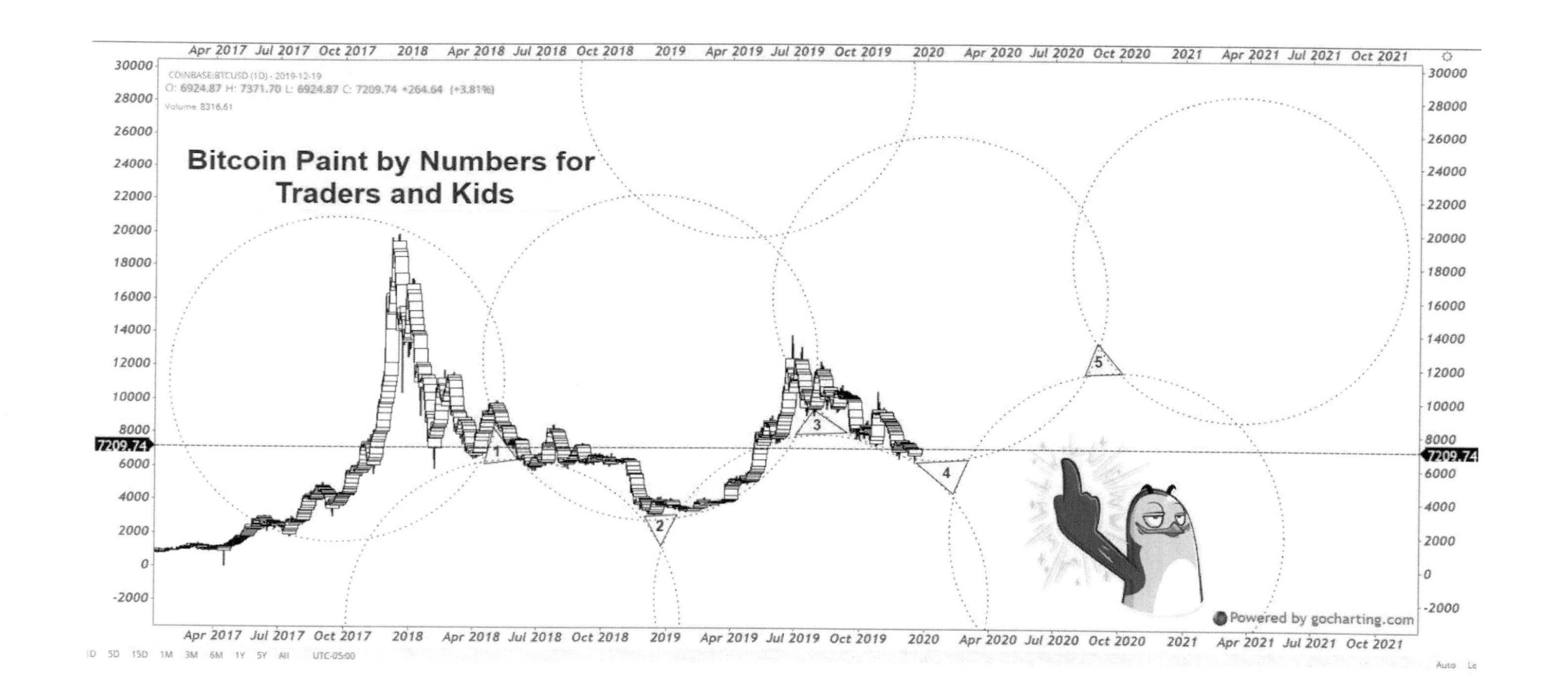

Bitcoin Paint by Numbers for Traders and Kids
COINBASE:BTCUSD (1D) - 2019-12-19
O: 6924.87 H: 7371.70 L: 6924.87 C: 7209.74 +264.64 (+3.81%)
Volume 8316.61
Apr 2017 Jul 2017 Oct 2017 2018 Apr 2018 Jul 2018 Oct 2018 2019 Apr 2019 Jul 2019 Oct 2019 2020 Apr 2020 Jul 2020 Oct 2020 2021 Apr 2021 Jul 2021 Oct 2021
30000 28000 26000 24000 22000 20000 18000 16000 14000 12000 10000 8000 6000 4000 2000 0 -2000
7209.74
1
2
3
4
5
Powered by gocharting.com
1D 5D 15D 1M 3M 6M 1Y 5Y All UTC-05:00
Auto

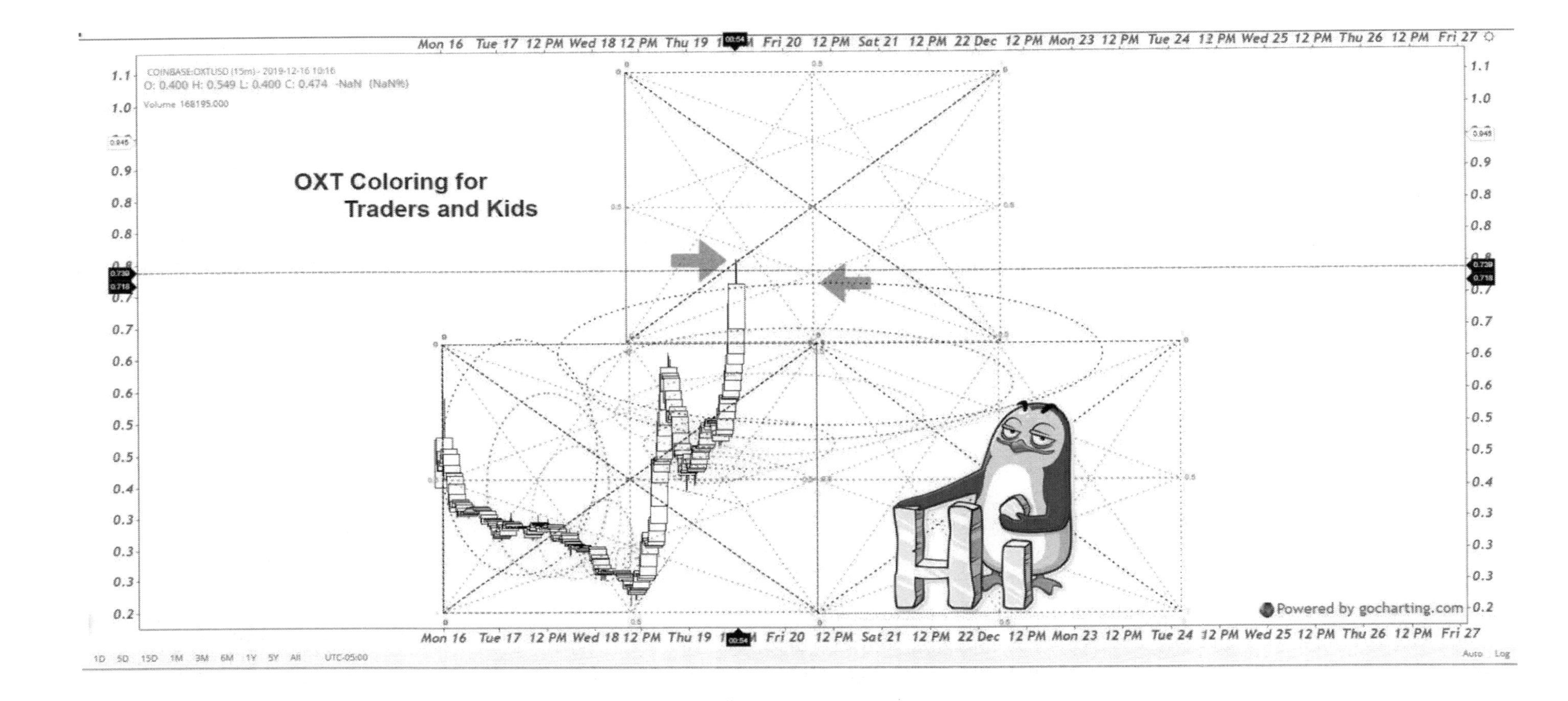
COINBASE:OXTUSD (15m) - 2019-12-16 10:16
O: 0.400 H: 0.549 L: 0.400 C: 0.474 -NaN (NaN%)
Volume 168195.000
OXT Coloring for
Traders and Kids
Mon 16 Tue 17 12 PM Wed 18 12 PM Thu 19 Fri 20 12 PM Sat 21 12 PM 22 Dec 12 PM Mon 23 12 PM Tue 24 12 PM Wed 25 12 PM Thu 26 12 PM Fri 27
HI
Powered by gocharting.com
1D 5D 15D 1M 3M 6M 1Y 5Y All UTC-05:00
Auto Log

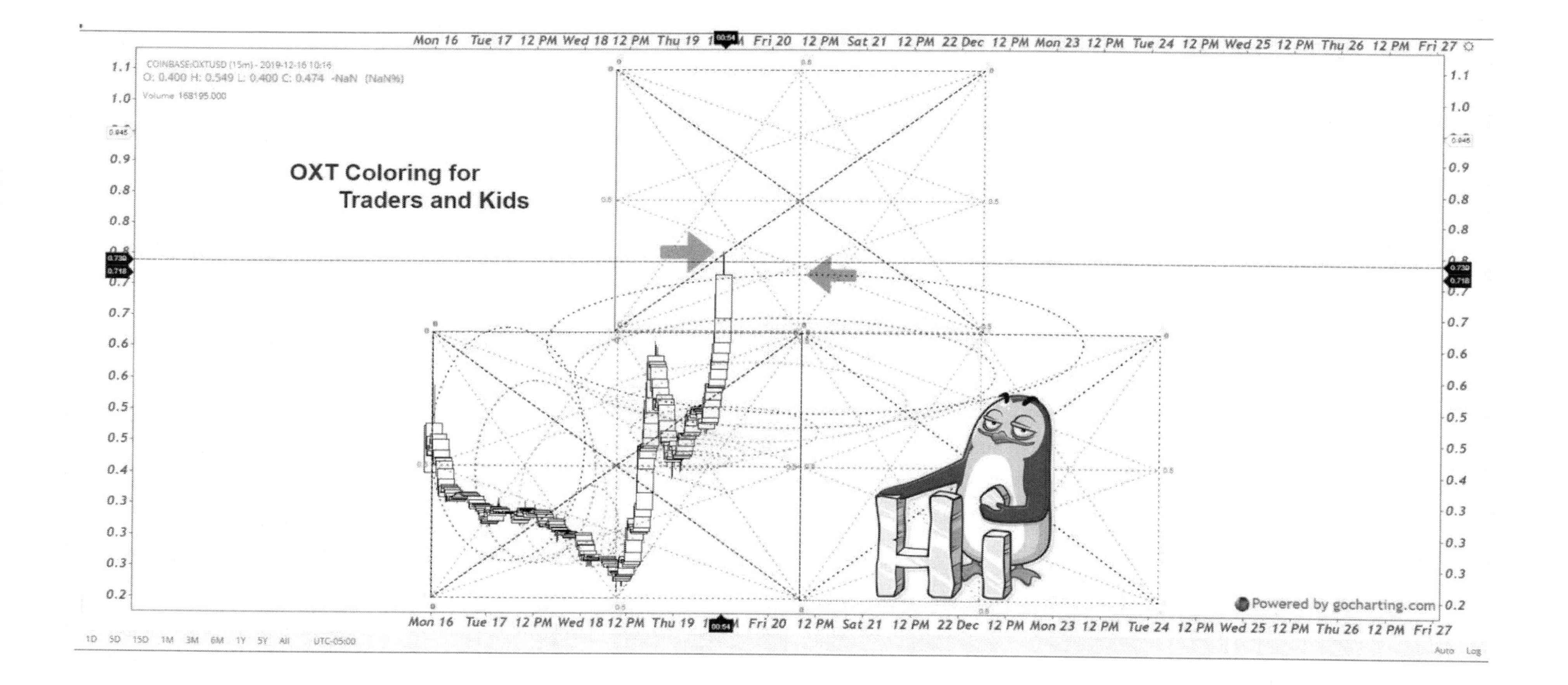
COINBASE:OXTUSD (15m) - 2019-12-16 10:16
O: 0.400 H: 0.549 L: 0.400 C: 0.474 -NaN (NaN%)
Volume 168195.000
OXT Coloring for
Traders and Kids
Mon 16 Tue 17 12 PM Wed 18 12 PM Thu 19 Fri 20 12 PM Sat 21 12 PM 22 Dec 12 PM Mon 23 12 PM Tue 24 12 PM Wed 25 12 PM Thu 26 12 PM Fri 27
Powered by gocharting.com
1D 5D 15D 1M 3M 6M 1Y 5Y All UTC-05:00
Auto Log

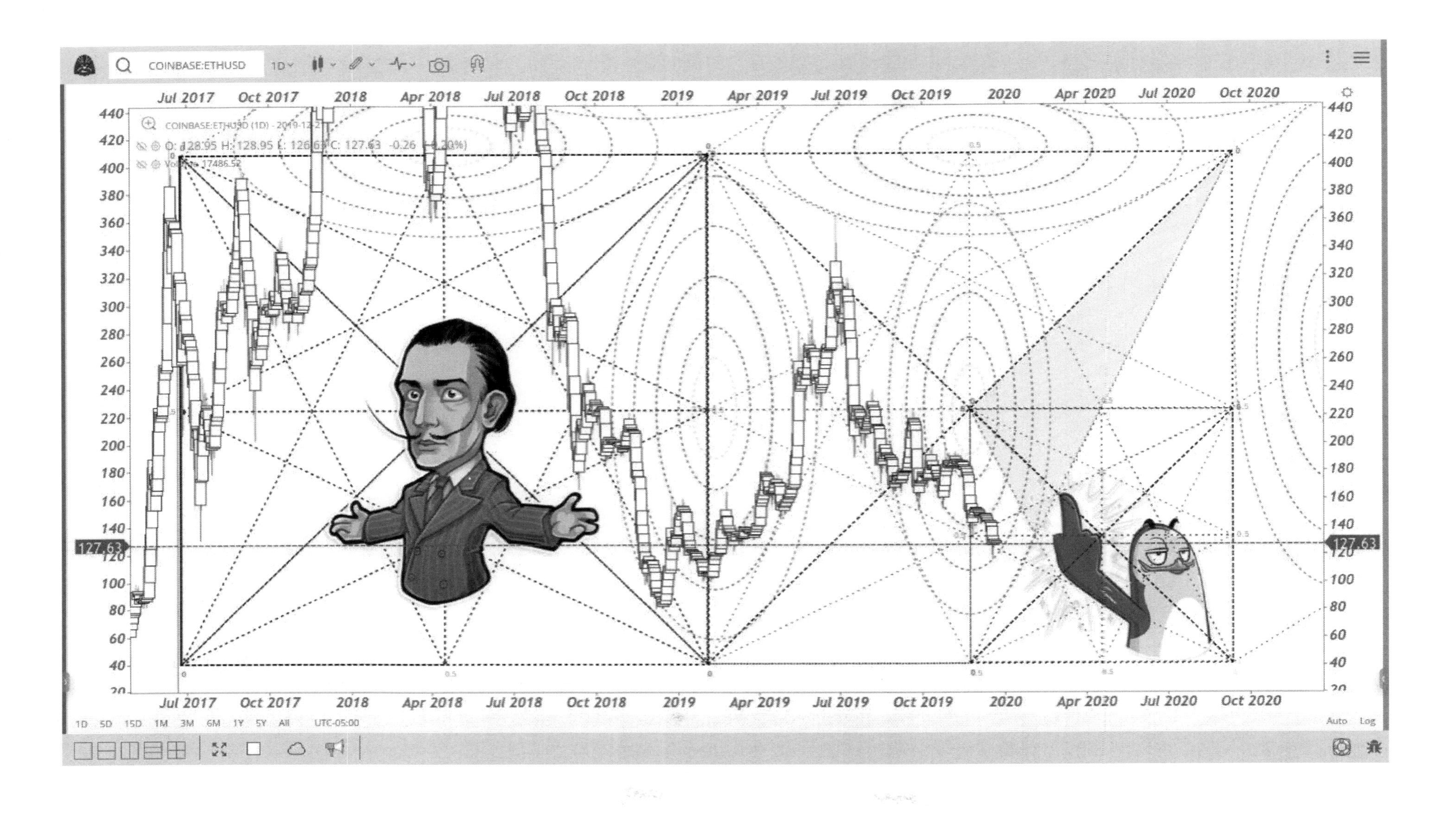
COINBASE:ETHUSD
1D
Jul 2017
Oct 2017
2018
Apr 2018
Jul 2018
Oct 2018
2019
Apr 2019
Jul 2019
Oct 2019
2020
Apr 2020
Jul 2020
Oct 2020
440
420
400
380
360
340
320
300
280
260
240
220
200
180
160
140
127.63
120
100
80
60
40
1D 5D 15D 1M 3M 6M 1Y 5Y All
UTC-05:00
Auto
Log

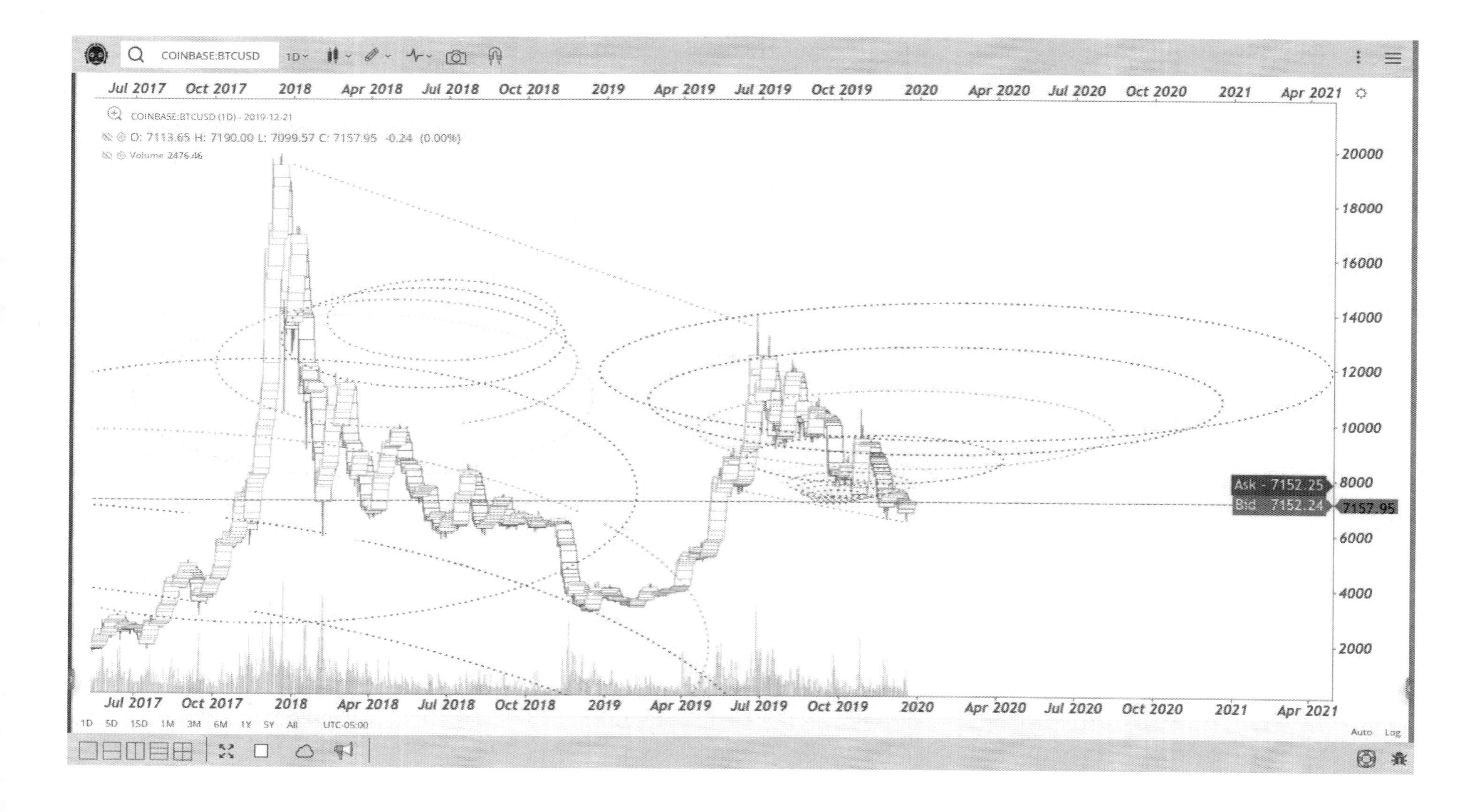
COINBASE:BTCUSD
1D
Jul 2017
Oct 2017
2018
Apr 2018
Jul 2018
Oct 2018
2019
Apr 2019
Jul 2019
Oct 2019
2020
Apr 2020
Jul 2020
Oct 2020
2021
Apr 2021
COINBASE:BTCUSD (1D) - 2019-12-21
O: 7113.65 H: 7190.00 L: 7099.57 C: 7157.95 -0.24 (0.00%)
Volume 2476.46
20000
18000
16000
14000
12000
10000
8000
7157.95
6000
4000
2000
Ask - 7152.25
Bid 7152.24
1D 5D 15D 1M 3M 6M 1Y 5Y All
UTC-05:00
Auto Log

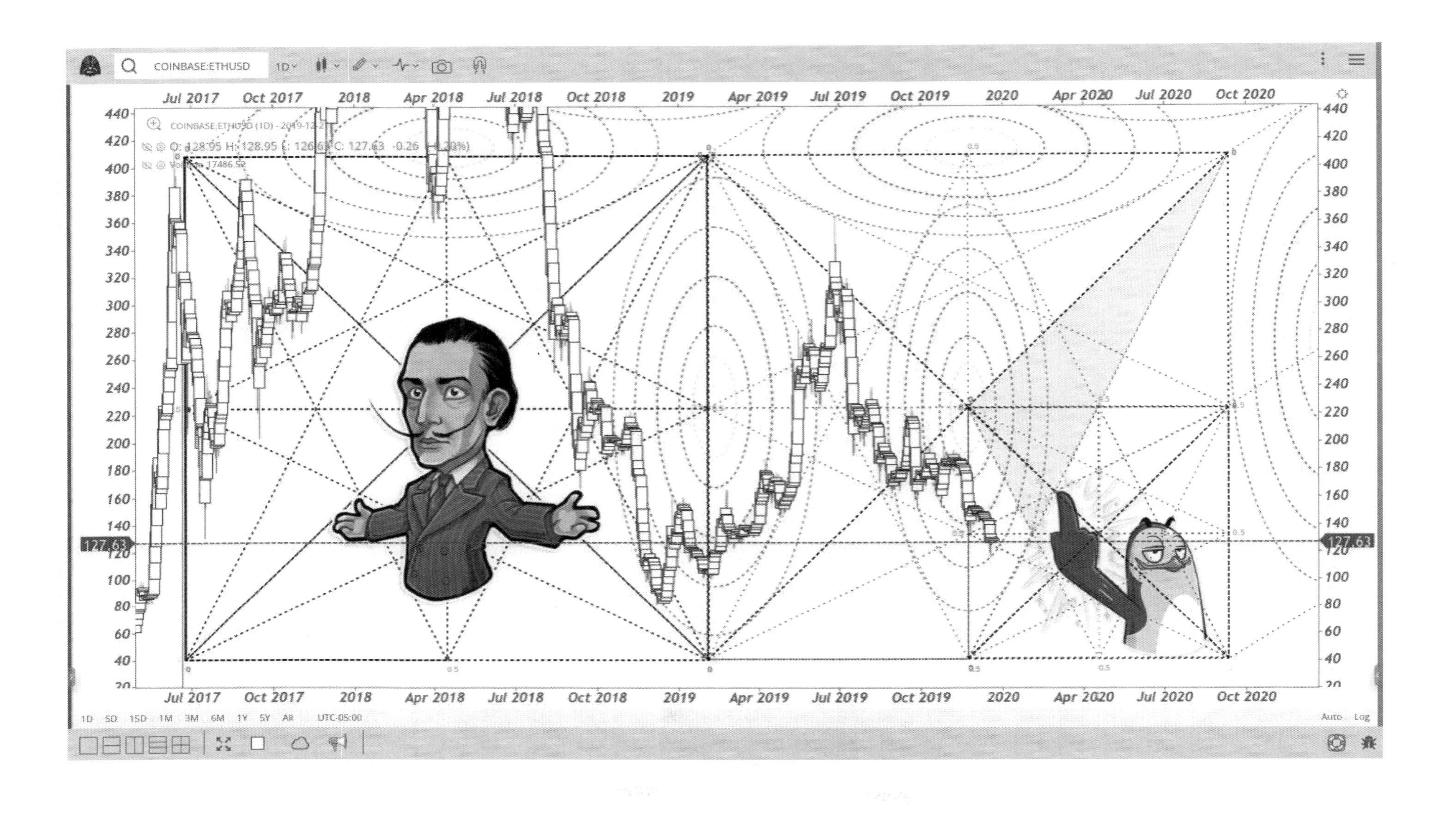
COINBASE:ETHUSD
1D
Jul 2017
Oct 2017
2018
Apr 2018
Jul 2018
Oct 2018
2019
Apr 2019
Jul 2019
Oct 2019
2020
Apr 2020
Jul 2020
Oct 2020
440
420
400
380
360
340
320
300
280
260
240
220
200
180
160
140
127.63
120
100
80
60
40
1D 5D 15D 1M 3M 6M 1Y 5Y All
UTC-05:00
Auto Log

COINBASE:ETHUSD
1D
Jul 2017
Oct 2017
2018
Apr 2018
Jul 2018
Oct 2018
2019
Apr 2019
Jul 2019
Oct 2019
2020
Apr 2020
Jul 2020
Oct 2020
440
420
400
380
360
340
320
300
280
260
240
220
200
180
160
140
120
100
80
60
40
127.63
1D 5D 15D 1M 3M 6M 1Y 5Y All
UTC-05:00
Auto
Log

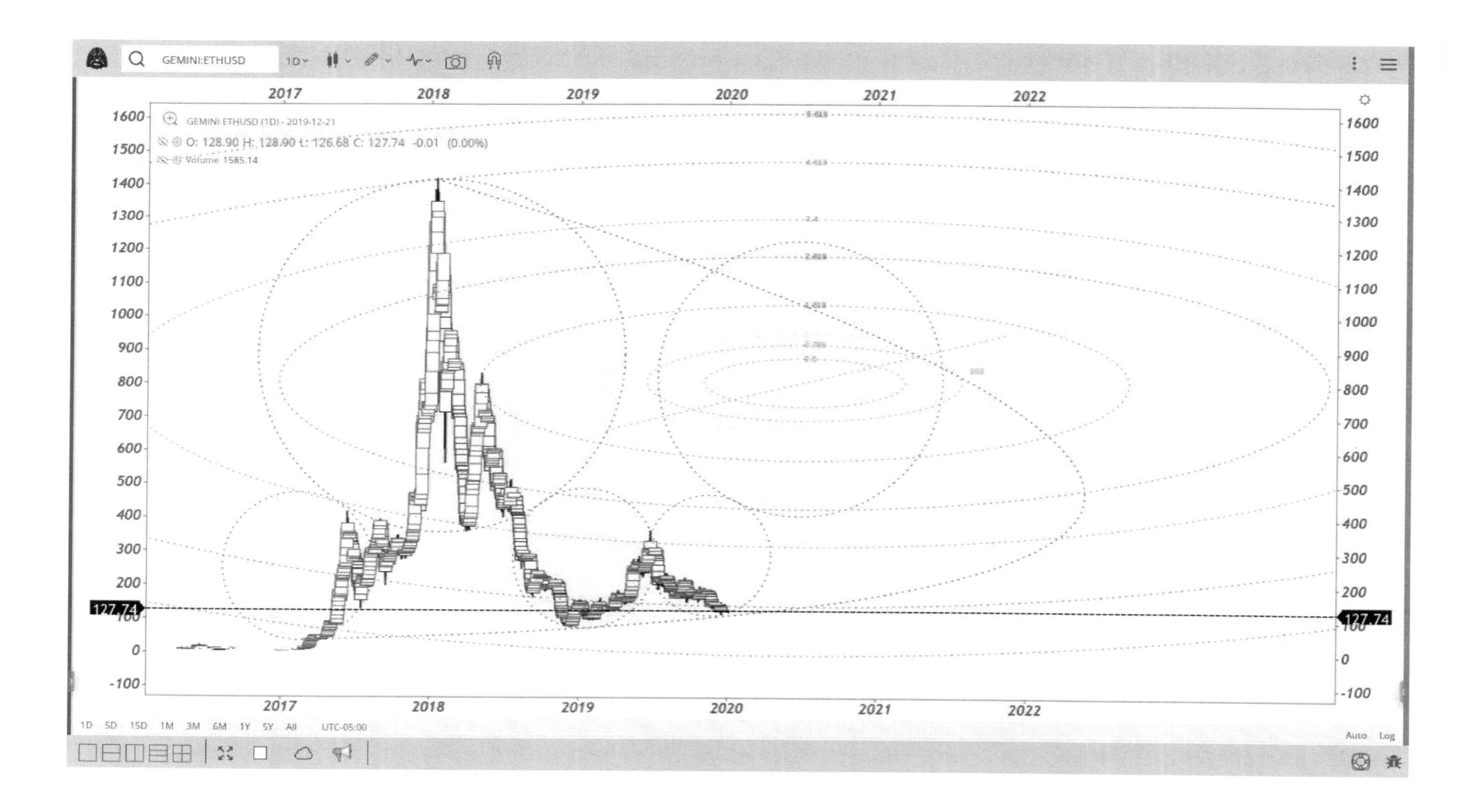
GEMINI:ETHUSD
1D
2017
2018
2019
2020
2021
2022
GEMINI:ETHUSD (1D) · 2019-12-21
O: 128.90 H: 128.90 L: 126.68 C: 127.74 -0.01 (0.00%)
Volume 1585.14
1600
1500
1400
1300
1200
1100
1000
900
800
700
600
500
400
300
200
127.74
100
0
-100
1D 5D 15D 1M 3M 6M 1Y 5Y All
UTC-05:00
Auto
Log

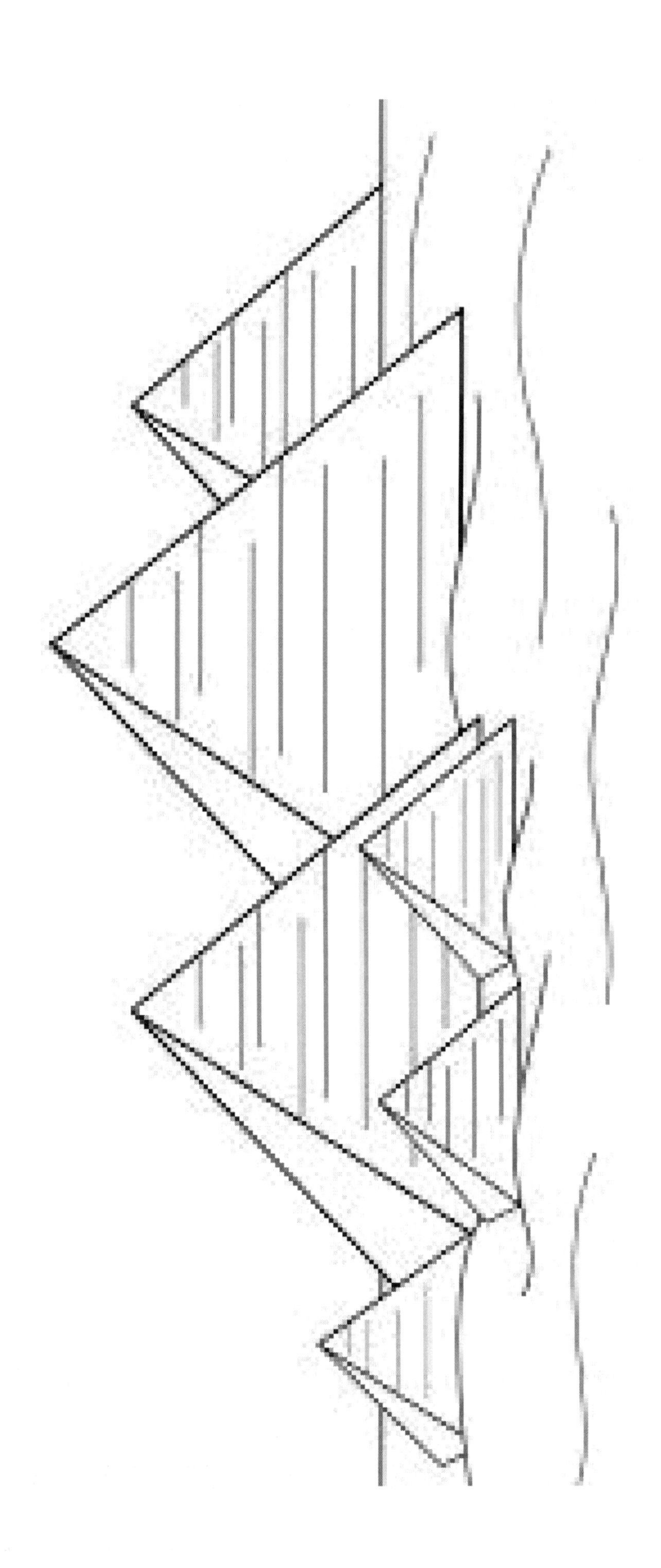

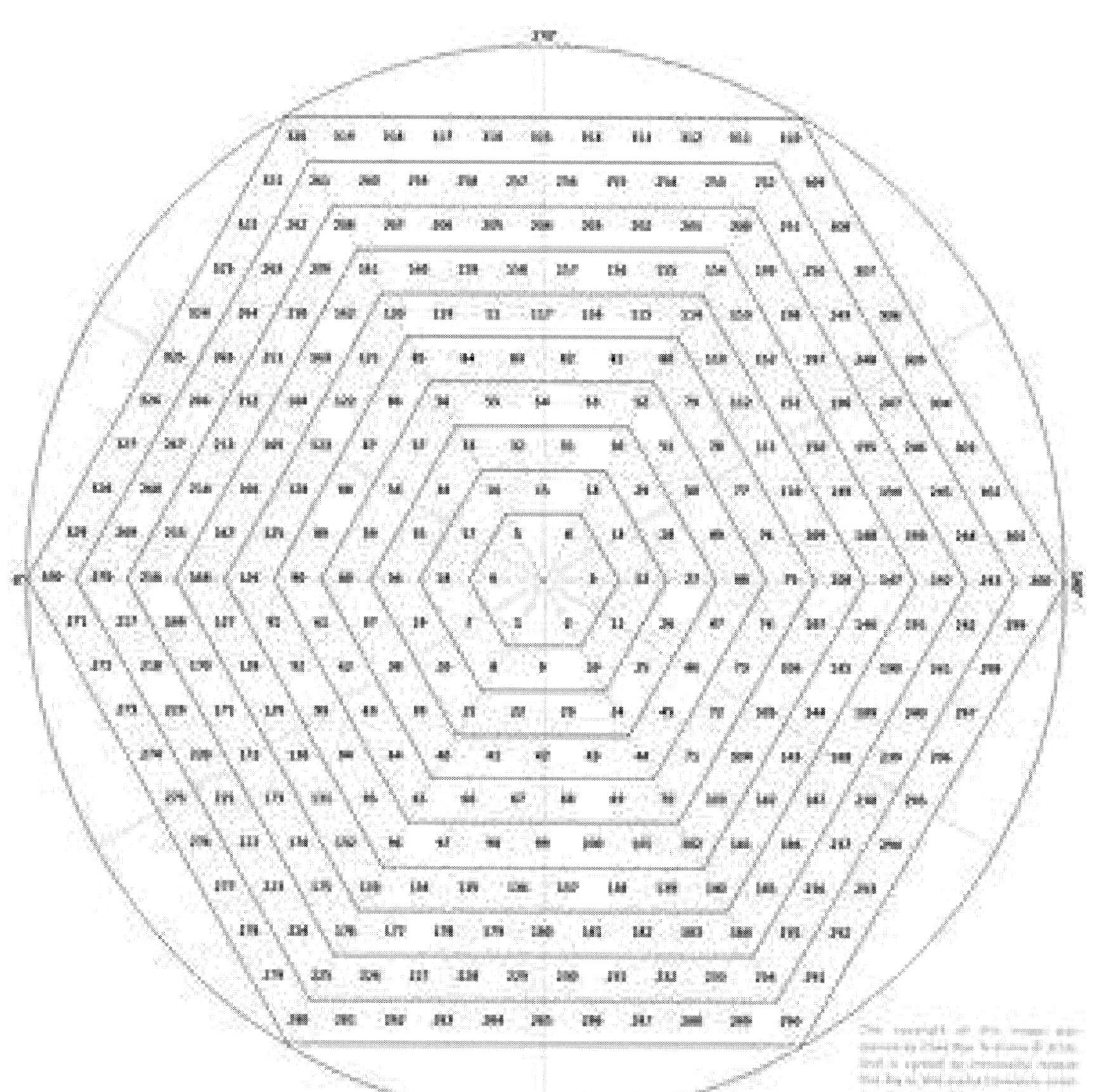

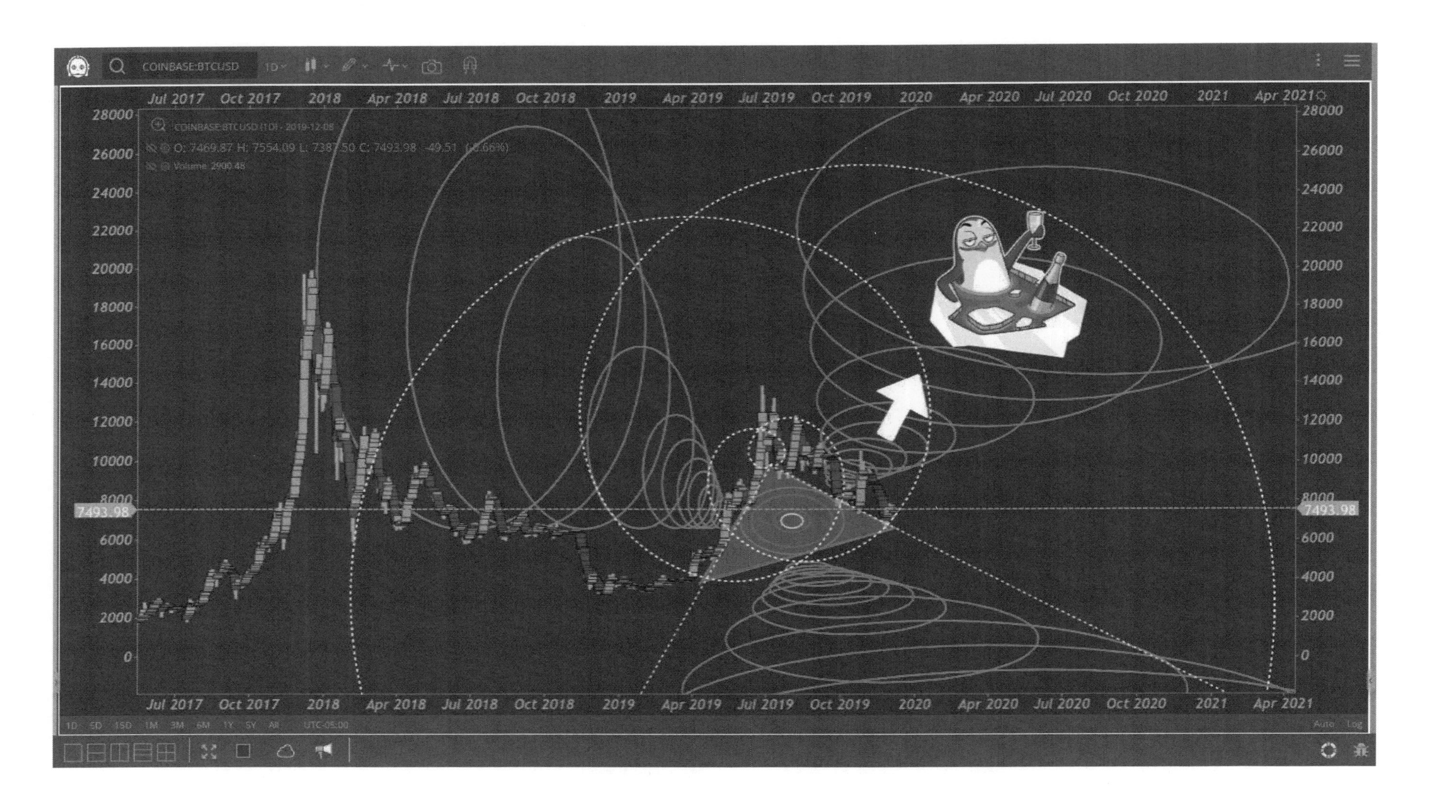
COINBASE:BTCUSD
COINBASE:BTCUSD (1D) · 2019-12-08
O: 7469.87 H: 7554.09 L: 7387.50 C: 7493.98 -49.51
Volume 2900.48
Jul 2017 Oct 2017 2018 Apr 2018 Jul 2018 Oct 2018 2019 Apr 2019 Jul 2019 Oct 2019 2020 Apr 2020 Jul 2020 Oct 2020 2021 Apr 2021
28000 26000 24000 22000 20000 18000 16000 14000 12000 10000 8000 6000 4000 2000 0
7493.98
1D 5D 15D 1M 3M 6M 1Y 5Y All UTC-05:00
Auto Log

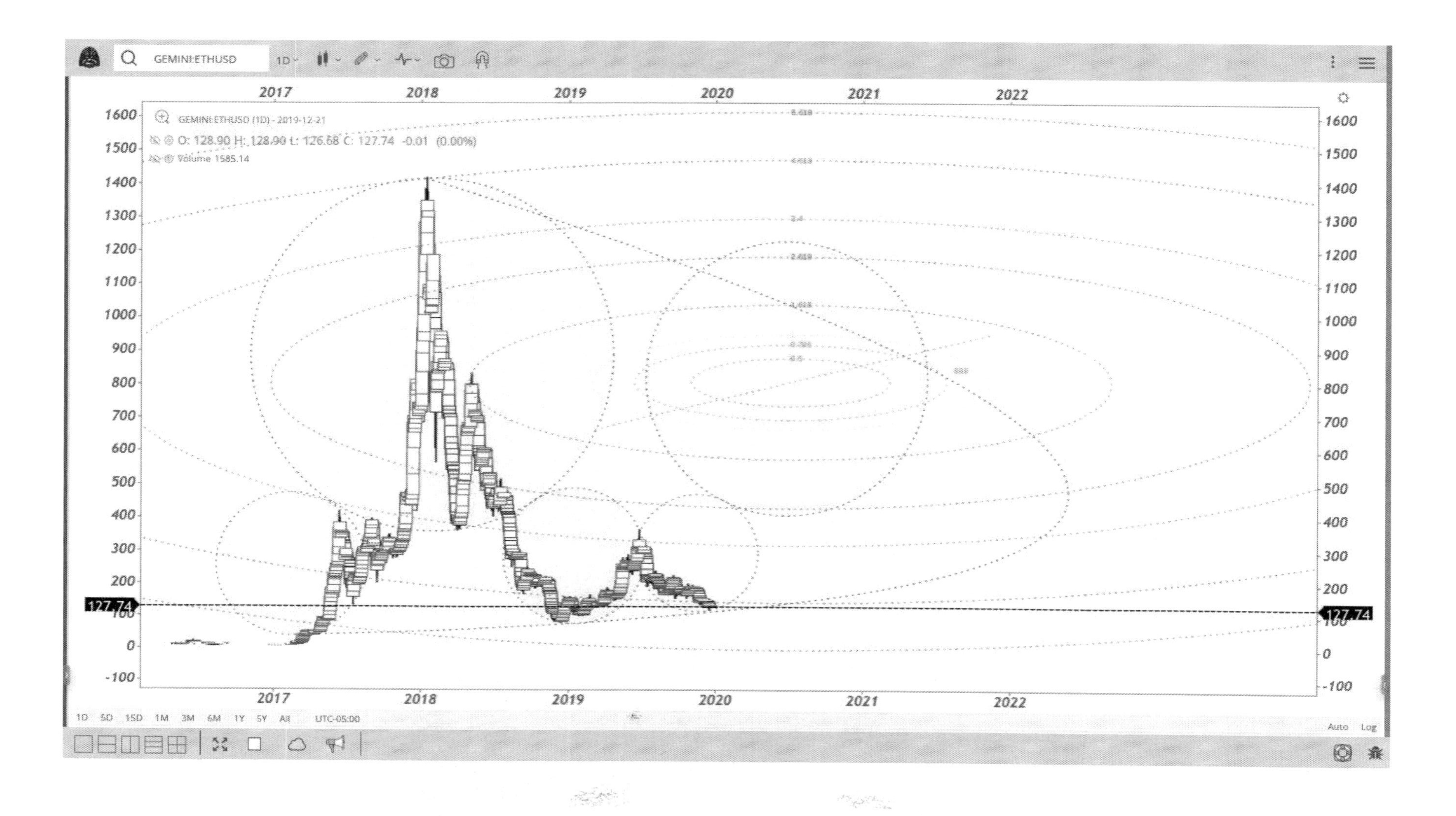
GEMINI:ETHUSD
1D
GEMINI:ETHUSD (1D) - 2019-12-21
O: 128.90 H: 128.90 L: 126.68 C: 127.74 -0.01 (0.00%)
Volume 1585.14
2017
2018
2019
2020
2021
2022
1600
1500
1400
1300
1200
1100
1000
900
800
700
600
500
400
300
200
127.74
100
0
-100
1D 5D 15D 1M 3M 6M 1Y 5Y All
UTC-05:00
Auto Log

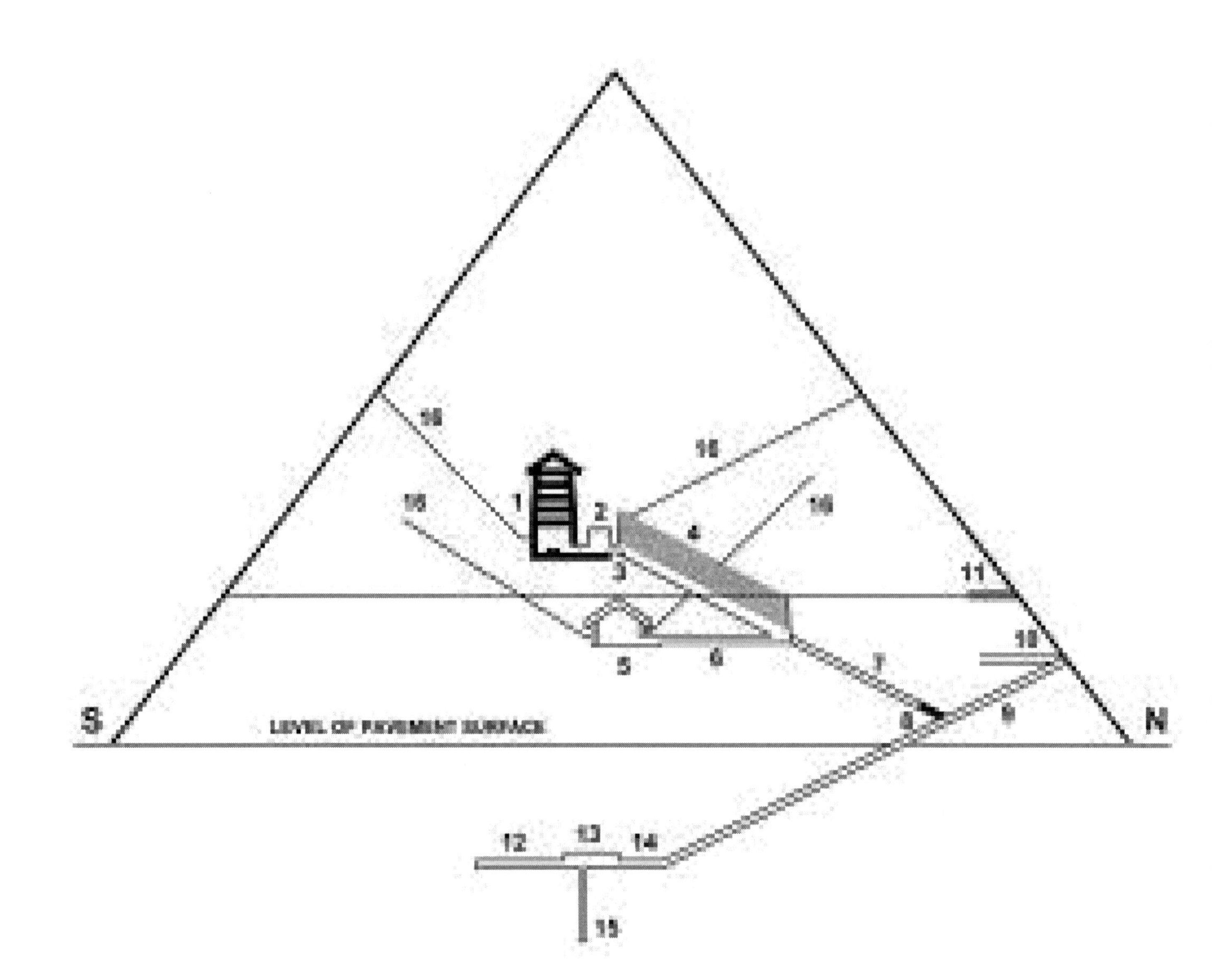
S
N
LEVEL OF PAVEMENT SURFACE
1
2
3
4
5
6
7
11
12
13
14
15

Color this BOOK 3 times

111

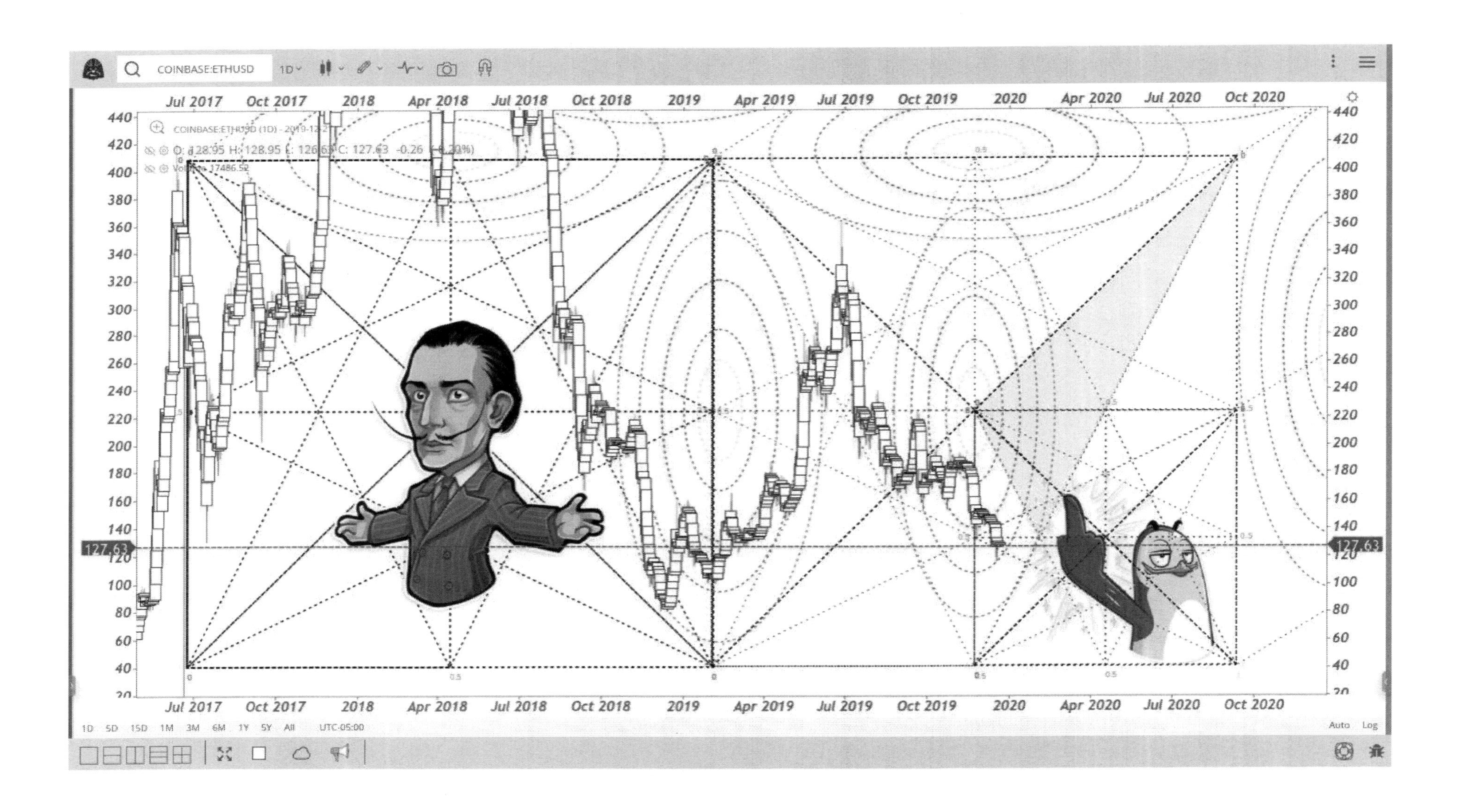
COINBASE:ETHUSD
1D
Jul 2017
Oct 2017
2018
Apr 2018
Jul 2018
Oct 2018
2019
Apr 2019
Jul 2019
Oct 2019
2020
Apr 2020
Jul 2020
Oct 2020
440
420
400
380
360
340
320
300
280
260
240
220
200
180
160
140
127.63
120
100
80
60
40
20
1D 5D 15D 1M 3M 6M 1Y 5Y All
UTC-05:00
Auto Log

This page intentionally left blank

Start over

1:47
69%
Candleboxai
26 members, 2 online
Will be sold for 144.14 USD
1:45 PM
I expect to sell zero but get a million views 😂😂😂
1:46 PM
Zero fucks
1:46 PM
GIF
-20 -10 -7 -5 -3 -2 -1 0 +1 +2 +3
Give-A-Fuck-O-Meter
1:46 PM
Pieter BX
GIF
Omfg that will be on page 25
1:46 PM
😂😂
1:46 PM
Pieter BX admin
😂😂
1:46 PM
Can I screenshot this and add as last p
pieter?
1:47 PM
Message

Made in the USA
Middletown, DE
07 September 2022

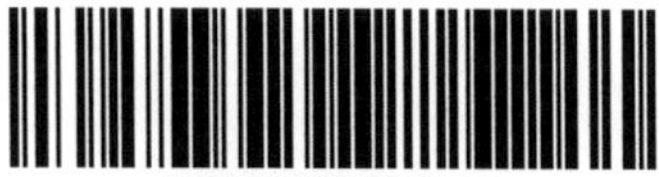